conjure back up in the digital present—and for that very reason it feels important to try.

In that analog studio, there was a feeling when the tape started rolling that this was the moment we would capture—a feeling of time moving both more slowly, and more quickly than usual.

Like when you're in an accident. Each split second is suddenly so palpable, as if you're living in slow motion. Yet what do we say when it's over? It all happened in an instant.

W9-AIX-662

Tascam reel-to-reel
tape machine

Galaxie 500, "Tugboat,"
7-inch vinyl, Aurora, 1988

Analog recording is like an accident in other ways. On tape, there was no "undo." You could try again, if you had the time and money. But you couldn't move backwards. What's done is done, for better and worse.

Today, life as a musician is very different. In the digital studio—and I am using one now— everything you do is provisional. That is, it can be redone, reshaped, rebuilt.

There's no commitment, because each element of a recording can be endlessly changed—it can even be conjured from digital scratch, as it were, and entered into a computer directly as data, without anyone performing at all.

This means there's no moment from lived experience that is captured forever—and unalterably so—in the digital studio.

Which is why it's more than nostalgia that makes me remember the analog studio as different than what we know today. Because the digital era has not just altered our tools for working with sound—or image, or moving images. It is changing our relationship to time itself.

[THEME: Robert Wyatt, "Trickle Down"]

This is WAYS OF HEARING. A six-part podcast from Radiotopia's *Showcase*, exploring the nature of listening in our digital world.

I'm Damon Krukowski, a musician and a writer. In each episode, we'll look at a different way that the switch from analog to digital audio is changing our perceptions: of time, of space, of love, money, and power. This is about sound, the medium we're sharing together now. But I'm worried about the quality of that sharing— because we don't seem to be listening to each other very well right now in the world. Our voices carry further than they ever did before, thanks to digital media. But how are they being heard?

[END THEME]

Contents

Time

Space

Love

Money

Power

Signal & Noise

Damon Krukowski
Ways of Hearing

The MIT Press
Cambridge, Massachusetts
London, England

This book was set in Univers and Univers Typewriter by James Goggin at Practise. Printed and bound at Wilco Art Books in the Netherlands.

Library of Congress Cataloging-in-Publication Data

Names: Krukowski, Damon, author.
Title: Ways of hearing / Damon Krukowski ; interlude by Emily Thompson.
Description: Cambridge, MA : MIT Press, [2019]
Identifiers: LCCN 2018028591 | ISBN 9780262039642 (paperback : alk. paper)
Subjects: LCSH: Music—Social aspects. | Music and technology. | Sound recording industry.
Classification: LCC ML3916 .K805 2019 | DDC 781.1/7—dc23 LC record available at https://lccn.loc.gov/2018028591

Episode 1

Time

Musicians know time is flexible.

Classical players call this *tempo rubato,* Italian for "stolen time." You steal a bit here, give a bit back there.

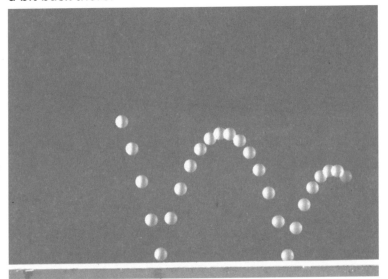

Jazz players call it swing.

Rock and funk players call it groove.

This flexibility of time is something we're all familiar with, whether or not we play an instrument.

There's the time it takes for dawn to arrive while you lay awake after a nightmare. And there's the time it takes when you're out all night with your friends.

What musicians' terms like *rubato,* swing, and groove acknowledge is that time is experienced, not counted like a clock. And our experience of time is variable—it's always changing.

Even if our eyes look at a clock and see twelve equal hours, our ears are ready to steal a bit here, give a bit back there.

And the analog media we use to reproduce sound—records and tapes—reflect this variable sense of time. A time that's elastic.

[SFX: Victrola]

The first turntables, Victrolas, were hand-cranked—you wound up a spring, which spun the platter as it wound down. The speed was hardly steady—at least never for long.

At the opposite end of the 20th century, hip hop DJs used speed controls on turntables to change the tempo of a record—to match it to another and keep the groove going. Or in the studio, to pile up samples from older records and make a new one.

[BED: A Tribe Called Quest, "Can I Kick It?"]

Victor IV gramophone, 1905

In 1990, A Tribe Called Quest sampled the bass from "Walk on the Wild Side"—along with the surface noise from the LP— added it to a drum loop from a Lonnie Smith LP, and made a new hit out of it.

They also had to give up on collecting any royalties.

Ali Shaheed Muhammad One of the biggest samples that we were able to clear was "Walk on the Wild Side" by Lou Reed. And by clearing that, he just took 100% of that song.

Frannie Kelley Wait. Really?

ASM He did.

FK You're saying Lou Reed gets 100% of the publishing?

That's Ali Shaheed Muhammad, from A Tribe Called Quest, and Frannie Kelley—they host a podcast about hip hop called *Microphone Check*.

ASM Yeah. That's his song.

DK Ali I always thought, you know my band Galaxie 500, we ripped off the Velvet Underground way more than you did! And we didn't have to pay them a dime.

I asked Ali how Tribe used the variable speed of turntables to make their own records.

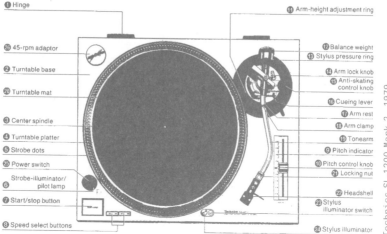

❶ Hinge

❷ 45-rpm adaptor

❷ Turntable base

❷ Turntable mat

❸ Center spindle

❹ Turntable platter

❺ Strobe dots

❷ Power switch

❻ Strobe-illuminator/ pilot lamp

❼ Start/stop button

❽ Speed select buttons

⓫ Arm-height adjustment ring

⓬ Balance weight

⓭ Stylus pressure ring

⓮ Arm lock knob

⓯ Anti-skating control knob

⓰ Cueing lever

⓱ Arm rest

⓲ Arm clamp

⓳ Tonearm

❾ Pitch indicator

❿ Pitch control knob

㉑ Locking nut

㉒ Headshell

㉓ Stylus illuminator switch

㉔ Stylus illuminator

Technics SL-1200 Mark 2, 1979

ASM You know sometimes we would change the speed on a turntable, because the staple hip hop turntable is the Technics 1200 — so that has a pitch control, and we would sometimes pitch it high there. Sometimes a 33 rpm record, you put it on 45. And one thing with Tribe — we layered certain sounds and so you kind of had to pitch them, cause often they weren't in the same key.

Layering these sounds, A Tribe Called Quest discovered something else about their samples.

ASM The records we were recording—they might not have been playing to a metronome, and they were free-flowing, so there are moments within, let's say, even a two-bar phrase, that maybe a drummer or bass player like were aligned in the first two bars and the tempo seemed to be consistent. And then the last three beats of a bar it sped up, or slowed down.

[BED: Galaxie 500, "Parking Lot"]

In 1988, the songs my bandmates and I recorded slowed down and sped up, just like Ali Shaheed described.

I know, because I was the drummer.

We played as steadily as we could—but this was a performance. We were nervous and excited. And we sped up at the chorus.

You might find that a flaw in our recordings. Or you might feel it's a part of their charm. Musicians have been speeding up at the chorus for as long, I'm sure, as there have been choruses. Our experience of time is *flexible*.

But not long after we recorded that first Galaxie 500 album, the commercial music world for the most part decided that this was most definitely a flaw. Bands on major labels—even bands who came out of our DIY, indie scene—started recording to a "click track." That is, a metronome.

Metronome

Fashions in music come and go—at the time, everyone was also using the swirling effect called a "flanger."

[POST: Intro to The Cars, "You're All I've Got Tonight"]

No one remembers why.

But the click track proved more than a fad—it was a change that stuck.

Because what truly changed in popular music during the 1980s was that digital machines entered the mix for the first time. And machines have a different sense of time.

Boss DB-60 Dr. Beat metronome, 2004

[BED: Human League, "Don't You Want Me,"
with LinnDrum beat]

The first drum machines to be widely used in commercial recordings were introduced in 1980. And MIDI—Musical Instrument Digital Interface, the language machines use to share musical information with one another—was launched in 1983. Together, these tools can lock musical time to a clock—which doesn't speed up at the chorus.

When Galaxie 500 recorded, we played our songs in what audio engineers now refer to as "real time."

Real time, as the name implies, is lived time—time as we experience it in the analog world.

Digital time is not lived time. It's machine time. It's locked to a clock. And that clock—a time*code*—makes everything more regular than lived time.

Logic Pro screenshot of the podcast audio corresponding to the text on this page

Time in the digital studio is stored in discrete cells on a grid, like figures on a spreadsheet.

And if you've ever worked with a spreadsheet, you know those cells can be sorted any which way. And refigured to most any desired outcome.

What this means is that a given experience of digital time is only one among many, equally plausible experiences.

Take this podcast you're listening to. Since this is a digital medium, you could choose to hurry me up—without changing the pitch of my voice, seemingly without changing any of my words or pronunciation.

Which means I can be

[SFX: 1.5× speed]

equally fast, speaking at 1.5 times my normal pace, like I'm manic.

[SFX: 0.75× speed]

Or, I can be equally slow, speaking at three-quarters normal pace, like I'm drunk.

Neither need be how this happened in "real time." In fact, none of this need ever have happened, in the sense that we understood that term in 1988.

[BED: Hatsune Miku]

That's a track made entirely by software—the singer is a program made by Yamaha, called Vocaloid.

Vocaloid tracks have been cropping up on the Japanese pop charts for a number of years, and are now making their way into other forms of music around the world.

As for podcasts… Adobe is developing a program called VoCo that will read text aloud for you—in your own voice.

What could possibly go wrong?

Making time conform to machines—making it regular—would seem to make it more unified. Like a standardized timeline that any of us can tap into at any moment.

But there's a surprising twist to digital time. It's actually very difficult to synchronize.

It's a challenge, in digital recording, to line up the different layers that make up a song. If you play a guitar into a digital recorder, and then play it back and add singing, your voice won't line up with the music exactly the way you performed it.

The reason is what's called, "latency."

Latency is the lag in digital communications introduced by the time it takes a computer to process them. Computers move very fast—that's what computers are good at. But analog time, it seems, moves faster.

Let's consider an example from outside music.

[POST: Castiglione home run call] *Deep down the line, way back! It is gone! A grand slam and the Red Sox have won it!*

That's the radio voice of the Boston Red Sox, Joe Castiglione.

Joe Castiglione, Boston
Red Sox announcer

Joe Castiglione I don't have a signature home run call, because all the good ones were taken. Mel Allen "going going gone," or "bye bye baby" Russ Hodges… At times I've used "forget about it," because you know it's gone. Some fans didn't like that, they said "we don't want to forget about it, it's a big home run!"

[AMBI: Archival play-by-play recordings]

JC Well the role of the broadcaster is to describe what you see and what is happening as it happens, that's why we broadcast in the present tense.

[POST: Castiglione calling pitches] *Roger deals and it's in there for a called strike.*

JC You can't be too soon, anticipate, cause you can get burned that way. Sometimes the ball doesn't carry as far as you think it's going to, or sometimes it carries farther. You can always tell if a broadcaster is behind the action, if the crowd is cheering and you're still building up to what's happening. You have to be quick, you have to be current; you have to be right on time.

Joe is such an important voice here in Boston, it used to be common practice to watch baseball games on TV, but with the sound off—and the radio on. So we could watch the game, but listen to Joe.

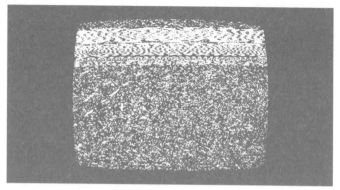
White noise on an analog TV

And then a strange thing happened. On June 12, 2009, Joe started calling the plays *before we saw them on TV.* That's because on that day, every TV station in the US switched from analog to digital transmission.

And that lag in the digital television image—the gap between it, and the real time of the game as Joe was calling it—that is latency.

It takes time for computers to translate the world into data. And translating data to analog so we can perceive it slows things down. Enough to put the tag after the call by Joe Castiglione.

[POST: Castiglione base running call]
Rounding third heading to the plate...Tucker's throw through in time and... he's out.

> JC There was an old saying in radio: "When you read it it's history, when you hear it it's news." And we are ahead of the satellite picture on TV, so we are first. And I think the immediacy of radio is something that is so critical and so germane to it.

Back before digital broadcast, that immediacy was audible in the city. You always knew when there was a home run at Fenway, because you heard a simultaneous, collective cheer all around Boston—from every open window and passing car.

[POST: Castiglione signature call] *Can you believe it?*

Today, those shouts are staggered. We get a splattering of cheers, like scatter points on a graph.

That's because latency is different—slightly different, but *different*—on each digital device receiving its own particular stream of digital information.

Professional music producers and engineers, in the digital studio, worry about latency all the time. The lag it introduces can be very very small. And yet... it's always there. And always threatening to knock the different layers of a song out of synch with one another.

If a young drummer today manages to cast a wide net around the beat, perhaps that's not so different from my own speeding up and slowing down on Galaxie 500 recordings. It may be a flaw, or it may be part of the charm.

Yet there's something very distinct about an experience of analog time—time that flexes, slower and quicker, *tempo rubato*—and this feeling of **blurred** time from latency.

For one, I can't think of a musical term for latency. Perhaps because it's not like anything we experience in lived time.

However, this experience of latency is typical of life *online.*

For all our seemingly instant and global digital communications, life online is in practice filled with chronological confusions. Social media like Facebook and Twitter shuffle posts according to various algorithms, most obviously popularity—the net result being that yesterday's news bulletin can interrupt today's string of think pieces about it.

Texting, too, is often tripped up by crossed posts, or missed posts, or abrupt silences followed by ten bings in a row—temporal hiccups in information exchange that would never happen face-to-face.

Hello?

Hello?

Hello?

Hello?

Texting in Apple's
iOS Messages app

In conversation—in real, analog time—we can gauge reaction to our statements before anything further is said. We can pause, without lapsing into silence. And we can fall silent, without ending the conversation.

[SILENCE]

Digital time is time designed for machines. We benefit in all kinds of ways from the convenience that makes possible.

But those conveniences come at a sacrifice.

Because when we trade broadcast for podcast, we give up the opportunity to experience time together—in the same instant—through our media.

And when we trade real time for machine time in music, we lose the ability to share our individual timing with one another. Musicians use *rubato*, swing, and groove to move us together—not locked to a machine, but to one another.

Analog time is flexible, but unified. We share it so easily, without tinkering with technology, without even thinking about it. Maybe that's why we seem to be surrendering it so quickly.

But without it, it's harder to share a moment in time together.

Like those moments I once shared with my band-mates in a recording studio—in a particular place, at a particular time—on West Broadway, just south of Canal, in New York City, 1988.

Thomas Struth, *West Broadway, Tribeca, New York*, 1978

Emily Thompson

Interlude

[SFX: Radio static resolving into a clear
signal, as when tuning into a strong station]

Radio Announcer We interrupt our regularly
scheduled program to bring you this special
announcement.

This is Emily Thompson. I'm a historian, and I'm coming
to you not from West Broadway in 1988, but from my
dining room in Princeton, New Jersey, thirty years later.
It's a wet November day and I can hear the birds at my
backyard feeder through the window to my left. Coming
from the right are the sounds of my neighbor's garage
being torn down across the street. From my kitchen, the
refrigerator hums and a radio faintly adds some classical
piano to the mix. I'm usually tuned to WNYC, a signal
that originates about fifty miles from here, on Varick
Street in New York City and not far from where Damon
just left you. But analog radio depends on the weather,
and the weather today is impeding my reception so I'm
listening to WWFM, from West Windsor, New Jersey,
which is just down the road. I could stream WNYC digitally, via my laptop, but I prefer my radio to travel through
the air, not through fiber optic cables. I like the fact
that what I hear depends on the weather between me and
the people making the signal; it connects me to them

through the physical space between us—be it fair or foul—and I enjoy turning the dial each morning to discover what I can pull in. It's a small sonic adventure whose outcome is unpredictable, and over which I hold only partial control.

Still, digitally delivered audio does have its place, and I particularly appreciate the ways it has enabled the sonic artform of the podcast to flourish. Podcasting is a medium that draws upon radio's rich heritage of storytelling and idea sharing, while accommodating most modern listeners' desire to be fully in control, to set the time and place and substance of their audio engagement. A download is thus more personal than a broadcast. Indeed, when heard through earbuds (as they typically are), a podcast can feel like an intimate conversation between creator and recipient, allowing each individual listener to imagine that it was created just for them. In this way, perhaps a podcast is also like a book. But unlike an audio book, which is simply a recording of a book being read out loud, a podcast originates in sound; it is designed to be heard.

One of the best podcasts I've heard is *Ways of Hearing*. I may be biased since I was invited to participate in its creation, but what I like most is not my own small contribution to Episode 2, but rather the cumulative effect that builds across all six episodes. That effect, simply put, is to challenge the fundamental premise of digital sound itself as a communication medium. Damon's podcast offers a McLuhanesque reminder for our post-analog age that the medium is still the message. Or rather, the message is: be aware of what your medium is up to, because it may not be telling you the whole story. And if that medium is being sold as "new and improved!" be very wary.

Through stories and conversations, *Ways of Hearing* provokes its listeners to examine carefully not only what we hear, but the ways that we listen. Damon reminds us that our senses' most fundamental task is not simply to entertain or distract us, but to situate our selves within our surroundings. By enabling each of us to understand where we are, our senses allow us to understand who we are, as individuals located within networks of physical and social connections. Just as John Berger's popular television series *Ways of Seeing* showed viewers back in the 1970s how to understand art—and the vision that art materializes—as a tool for locating our selves in our world, the *Ways of Hearing* podcast enunciates how temporal, spatial, and social meanings are all encoded in the sounds we hear. Or were, that is, until the sound itself was digitally encoded and all those meanings were discarded as noise, unworthy of algorithmic transformation. Through its own digital sounds, the *Ways of Hearing* podcast paradoxically calls attention to the things we no longer hear when we listen to digital content. It's an ode to noise, composed in the language of pure signal, the pristine and lossless bytes of perfectly reproducible ones and zeros.

Only someone as variously gifted, hard-working, and hard-thinking as Damon Krukowski could have pulled this off. He is a musician equally attuned to the art, the technology, and the economics of music making. He is a thinker whose analytics were forged in the crucible of grad school but who escaped that world before the smoke of those fiery furnaces could obscure the clarity of his ideas. And he is a publisher who understands the complex relationships between media forms and their content.

All these different areas of mastery are evident in the *Ways of Hearing* podcast. It's a heady mix of scholarly thoughts and personal memories, where a cast of characters that includes John Cage, Dave Grohl, and Damon's mom all contribute to an organic conversation that moves from the malleability of time to the nature of neighborhoods to the questionable proposition that sound is property. Voice, music, and sound effects are scored virtuosically, taking the listener on a tour of the digital soundscape while always seeking to answer the question, "How are these new sounds being heard?"

So, why now has that podcast been converted into this book? Why silence all those sounds by turning them into words printed on a page?

While clearly inspired by the book *Ways of Seeing*, John Berger's memorialization of his TV series between tactile covers, Damon's decision to publish *Ways of Hearing* as a book is not based in material fetishism. Rather, by deploying two different channels—podcast and print— to deliver his message, he intensifies its effect. Just as two almost-identical flat images, when viewed through a stereograph, suddenly transform into a three-dimensional panorama, or when two really-quite-similar monophonic recordings are played simultaneously through stereo speakers and the sound magically spreads out before you, Damon offers his message through two distinct sensory media that combine to form something bigger and better than both. The result not only deepens his message, but also locates it more fully in the social world that he urges us to embrace.

Let's assume that you've already listened to the podcast—I hope you have! As you listened, you probably imagined the people and places you heard, creating visuals in your mind to accompany the sounds in your ears. These imaginary pictures worked in conjunction with your earbuds—and I'm assuming you listened through earbuds, or perhaps you were entombed within the giant earbud that is a car-while-commuting—effectively removing you from the place where you actually were. That place—where you actually were—was just so much sensory noise (traffic, the unwanted press of subway seatmates, the clatter of a busy café) competing with the message that Damon brought. You filtered out all that noise to focus on the message. But that message was to hear the noise, to always be present in the place where you are in all its complexity. So where did that leave you? The podcast had created a paradox, its score ending with a suspended chord that may have left you, the listener, feeling unsatisfyingly unresolved.

Here now, with this book in your hands, you might once again be hoping to escape from your immediate surroundings, but the sensory modes are reversed. As you read and scrutinize the pictures of all the people and places described, you imagine the sounds that are now only evoked on the page. Assuming you aren't simultaneously listening to something else through your earbuds—if you are, take them out!—the sounds of the world you actually inhabit right now are mixing and mingling with the sounds that Damon has planted in your mind's ear. Your attention might move from one to the other, then back again, as you read. You look up if someone calls your name. You slide to the right if you hear two people conversing as they enter the subway car so that they can sit together. Through these small gestures, you act out the message that Damon delivers and resolve the paradox produced by the podcast. You can reclaim the lost noise of the digital soundscape and maintain your place in the social networks that sustain us all.

The external noises that necessarily accompany the silent signal of this book thereby enhance the message that the podcast had previously announced. One mode isn't really better than the other, and each still works pretty well on its own, since their messages are virtually identical. But the slight differences between them make all the difference. By calling our attention to the sounds our digital world has left behind, and then by providing a means to rediscover them, Damon has demonstrated how our world is enriched by noise.

In my world, back at my dining room table on this particular day, the rain has picked up and the birds have decamped to wherever it is birds go to wait out a downpour. The din of demolition across the street has ceased, as the structure has now been reduced to rubble. The fridge still hums, and the thump and gurgle of a sump pump now intermittently call up from the basement below. Amidst this gentle soundscape, the music on my radio rises to the fore, graduating from noise to signal. The announcer breaks in to identify the concerto and I realize I need to break away, to return you to the place that you are, on this day, and leave you there with Damon's ideas in hand. It's a very good place to be.

[SFX: A mechanical click, accompanied by a short burst of amplified sound, such as you get when you turn off the radio]

2

Elevator buttons in a building with no designated thirteenth floor

[SFX: Elevator bing]

I grew up on the fifteenth floor of an apartment building in Manhattan. Actually it was the fourteenth floor—there was no thirteen button in the elevator, to stave off bad luck.

It was an old building, and my family kept the windows open year round—even in winter, because the steam heat could be overpowering. Which means I heard the roar of the city at all hours, and in all seasons—an ocean-like white noise. A noise I missed whenever I tried to sleep in the countryside.

But other than that roar, it was really pretty quiet. Only a few individual sounds reach an apartment that high off the ground: sirens, honks, whistles, the occasional dog bark. Sounds of warning— sounds designed to rise above the rest.

As soon as I took the elevator down and walked outside, though, I was immediately swept up into that ocean of sound.

34th Street and 7th Avenue
facing north, New York, 1977

Up close, it was less like a roar and more like being hit by a massive wave—of sounds, and of people.

As a kid, I prided myself on being able to move quickly through those New York crowds. It was like a city version of surfing—anticipating how this wave of people and cars would move, getting in a groove, and riding it through.

Now I find the sidewalks in New York harder to navigate. I'm older, for sure. But I don't think my city instincts are so stale. I've spent a lot of my adult life on tour as a musician, traveling in cities—some twice the size of New York.

[SFX: Tokyo subway]

But the crowds in New York don't seem to move as quickly as they used to. It's like there's a lag in people's movements—a hesitation, where I expect action.

　　And it's no wonder, because if you're busy talking on the phone, or looking at a screen—are you really *in* this crowd?

　　Are you here, with me, on this sidewalk— moving together like we're all in the same wave?

　　Or are you in some other space altogether?

Manners poster, Metro Cultural Foundation, Tokyo Metro, 2013

In Tokyo, people on crowded trains pretend they're asleep, to avoid eye contact.

　　But here? With all these headphones, it's like we're avoiding *ear* contact.

Space

Listening has a lot to do with how we navigate space. We use our stereo hearing to locate sounds around us—and to map where we are, in relation to the source of those sounds.

If you stop up your ears—say, with earbuds listening to this podcast—you'll find you aren't as aware of the space around you. Or of other people. If you're on the street, you won't hear their footsteps approaching; you won't hear their cough letting you know they are right behind you; you may not even hear them yelling at you to get out of the way.

And if you're in New York, that person yelling might be Jeremiah Moss.

New York City crosswalk symbol,
Michael Bierut/Pentagram, 2012

Jeremiah Moss When I see people walking around with the screens, I see people who don't want to be here—they're not here. They're opting out of the street life of the city. So you're creating a private kind of bubble in which to move through public space. So in some ways the public has been triumphed over by the private. These little private bubbles that just bounce off each other.

Jeremiah Moss writes a blog called *Vanishing New York*. He's been observing the changes in the city since the 1990s. And he's angry.

> **JM** I walk around in a rage a lot of the time.

What's making Jeremiah angry is gentrification— the increasing wealth, and homogeneity, of the city. Jeremiah actually calls what he's been witnessing in New York *hyper*-gentrification, because it's moving so fast. And seems so out of control.

> **JM** I just want to point out this building, it has a car elevator so you can bring your car into your condo. Now if that's not suburban, I don't know what is. That is a vertical suburbia.

I met Jeremiah in Astor Place, a square that's always been a gateway to the East Village. When I was growing up, it was known as a place you could find things on the black market—drug deals were happening, just out of sight. And in plain view was a kind of thieves market—the contents of tourist suitcases, car

trunks, burglarized apartments, and simply old aban-
doned junk, all spread out on the sidewalk for sale.

> JM There's nothing to look at in this glass tower,
> you know—smooth slick surface—if you are on
> a screen, you can sort of glide right by it. And they
> look like screens. And what people do is look up
> from their screen going by this glass to look at
> themselves reflected in the glass. And then they
> go back to the screen.

Alamo Sculpture with Flea Market, Astor Place
to Cooper Square, early 1980s

Today, Astor Place is what's known by the city authorities
as a "pedestrian plaza." One of the streets that used to
cut through it has been closed off. There are tables with
umbrellas, and chairs. And it's surrounded by new glass
towers, with chainstores on the ground level: there's a
K-Mart, a Walgreens, *and* a CVS.
  Which means there are still a lot of drugs being
sold here.

Astor Place, 2016

JM What they're doing is they're kind of privatizing public space in this very stealth kind of way. So this is still public, everybody's welcome here. And yet… You'll see these private security guards walking around. This is a place you know where people would protest, there'd be public dissent in this space. And now you have these signs you know no skateboarding no this no that. Because it's now a pedestrian plaza, you can have rules—you can dictate how people use the space.

Walking through the East Village with Jeremiah Moss is like experiencing two urban spaces at once. There's the one we're in now, in hyper-gentrifying New York. And then there's the one that's underneath it, a ghost city of crowded tenements and streets that moved to a different rhythm.

[SFX: 1970s NYC street sounds]

New York City's Environmental Protection Administrator Jerome Kretchmer and a van for measuring city noise levels, 1971

It's a rhythm I knew well—from growing up in New York in the 1970s. And from being in a band who played there in the 1980s.

> **JM** Shall we go into what used to be CBGB's and is now John Varvatos...?

> **DK** Yes let's. Let's do it. I haven't dared walk in since it was CB's.

Jeremiah took me to what used to be the iconic rock club, CBGB's. It's now a boutique.

> **DK** Oh my God. So there's a kind of a fake stage here with a display of shoes and T-shirts for sale but also amps and guitars and drums. But they're obviously not being

played. I'm standing right where the stage was. They've left the graffiti on the walls stripped back a little bit. But as you can hear, the music playing is nothing like what this room represented.

[BED: Galaxie 500, "Don't Let Our Youth Go To Waste," live at CBGB's, 1989]

Playing gigs at CBGB's meant a lot to me and my bandmates in Galaxie 500, not because it was such an ideal space for music—it didn't have the best equipment, people had to walk past the stage to get to the

CBGB OMFUG
315 Bowery (at Bleecker) NYC (212) 982-4052

WED. OCT. 17
THE OUTSIDER·THRILL 13
THE CHILDREN · CALAMITY JANE
RED HERRING

THURS. OCT. 18 $10.
BLACK ROCK COALITION presents 5th ANNIVERSARY BENEFIT
DA DAH DOO DAH DA
DESTROY ALL MONSTERS
FAITH·ABSTRACT

FRI. OCT. 19 $10.
GALAXIE 500
ANTIETAM
CRYSTALIZED MOVEMENTS
LOVE CHILD

SAT. OCT. 20 $10.
fIREHOSE

dramatically filthy bathrooms, and there wasn't really anywhere you could leave your instruments safely.

But this was where so many great shows happened. The space had meaning because of the sounds that had been *shared* there.

When you got on that stage, you knew who else had been there.

CBGB's, The Bowery, 1986

Of course the city has always been something like this—a center of commerce and consumption, destroying whatever was for whatever will be next.

[SFX: 1920s NYC street sounds]

But if you talk to Emily Thompson, a historian at Princeton who knows an earlier New York as if it had never been buried under what came next—you start to see something more than a never-ending cycle of development. You start to see a timeline with a very definite story. One with a beginning, middle, and— if Jeremiah Moss is right—maybe an end, too.

For Emily Thompson, that story is about sound. And our efforts to control it.

> **Emily Thompson** The 1920s was a time that was perceived to be uniquely and unprecedentedly loud, and a lot of that had to do with technological changes in the modern city. The rise of internal combustion engines with vehicles everywhere, and you've got elevated trains running above the city streets, but cities like New York are simultaneously excavating to bury those trains and turn them into subways. So it just was hard for many people to deal with this new environment.

Postcard showing Fifth Avenue, 1927

Hard enough that the 1920s is when the decibel meter was invented. And civic groups put that new measurement of noise to use, lobbying for quiet zones around hospitals, rubber tires on cars, asphalt poured over cobblestone streets—all efforts to bring some kind of control over this new urban racket.

Acoustical engineers assess Times Square noise levels using a new sound measurement: the decibel, 1929

And yet—as any corner in New York still demonstrates today—those lobbying efforts didn't really succeed in controlling the sound in the street. But as Emily Thompson points out, they did lead to another, more effective set of changes.

> ET Looking at campaigns for noise abatement helped me understand the drive to control sound in interior spaces—to create kind of refuges from this environment.

A key place for experimenting with that kind of *interior* sound control was concert halls.

ET Maybe if I step here I can say I was on the stage.

DK It's so big, it's like a gymnasium.

Radio City Music Hall opened in 1932, and was a watershed in acoustical design. It was the first hall built for amplified sound.

It has a huge stage—big enough it once hosted a pro basketball game—because it doesn't matter how sound projects *off* it. Everything that happens on that stage was intended to be mic'd—even the Rockettes' feet—and heard through speakers built into the architecture of the hall, rather than through the natural reverberation of its space.

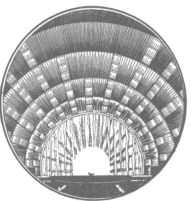

When nothing is being projected through those speakers, the room itself—despite its size—is actually very, very quiet. Emily and I visited Radio City together on a Saturday morning, when hardly anyone was in the building, and sat down to talk in two of its 6,000 empty seats.

Radio City Music Hall interior

But as you can hear, it sounded like we were in a tiny, carpeted recording studio.

ET Basically what we are talking about are the criteria for optimum reverberation—how much of a sound of space do you want. And I think that the ideal became that you were really just hearing that signal, the voices, the musicians that you wanted to hear, and everything else—the sound of the room in which you were experiencing this—became considered noise, something that was just interfering with what you really wanted to hear. And so auditoriums began to be designed to be more absorptive, to reduce reverberation. In the most extreme statement, some auditorium designers said well, walls are really just a necessary nuisance—they keep the weather out, they let you hang things off of them. But acoustically, an ideal auditorium would have no walls at all.

[SFX: Radio City's Wurlitzer organ]

Reducing reverberation to better control the sound *inside* a space ultimately led designers to imitate the sound of the *outdoors*, where there's no reverberation.

ET And here in Radio City, I think you can kind of approximate that with the luxury of such an enormous space to work with, and the highly absorptive surfaces out of which they built the walls—the arches, the back panels were all either covered with sound-absorbing cloth and fabric, or they used special acoustical plaster which absorbed the sound energy rather than reflected it back down into the audience. So what they were hearing was really what they had bought the ticket for—which was whatever was happening on stage, just kind of coming directly at them. And in the case of Radio City, actually via loudspeakers.

Curiously, no one much questioned this new sound of Radio City when it first opened. After endless debate surrounding the best acoustics for music—is it La Scala in Milan, is it Symphony Hall in Boston—Radio City, without *any* acoustics, was simply... accepted.

ET There were lots of other criticisms of the hall. People had problems with its massive size, they felt it was hard to see what was going on on stage. But people found the sound unremarkable, therefore I assume unproblematic. You know it was intended to fit in with the electroacoustic world of radio and then sound movies. So it's what people liked at the time, it's how they defined good sound. And I think that Radio City certainly delivered on that level when it opened.

March of Time newsreel screened in Radio City Music Hall, 1938

The earbuds we're adopting now—so universally, so quickly—also feel like a direct experience of the sounds they amplify. And just like the audience at the opening of Radio City, we're generally not questioning the quality of that sound.

On the contrary—based on the evidence all around us, the evidence Jeremiah Moss rails against—we obviously love it.

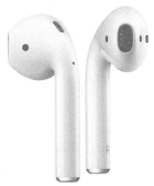

Apple AirPods, 2016

Earbuds are like a dream of the sound that Radio City's designers were after—they *are* an auditorium without walls. And our digital devices are not only portable, they tap into a seemingly limitless cloud of sounds.

[BED: Damon & Naomi, "New York City"]

We can now control and tailor our audio environment, to most whatever we desire it to be—in any space we find ourselves.

Yet many of those spaces—outside, in the street—remain as noisy as ever. The *Times* reports that New Yorkers are lodging twice as many noise complaints as they were just five years ago. For most, hyper-gentrification has only made our cities more difficult to navigate—with longer commutes from affordable neighborhoods, more people and cars at rush hour, more crowded public transportation.

Please don't
sound your
Horn in
New York City

Signage, mayoral "Committee for a Quiet City" anti-noise campaign, 1956

As in the 1920s, we're using interior sound to create a refuge. But our digital devices have extended that interior space—into the street itself.

Through audio, we're privatizing our public spaces. Just like the redesign of Astor Place into a "pedestrian plaza."

⌐ ¬
∟ ⌐

Historians don't generally like to speculate about the present. But while we were at Radio City Music Hall, I asked Emily Thompson to consider this 21st-century soundscape of earbuds.

ET Headphones are different for me though, because the sound is inside your head. It's not even outside of your head in the outdoor space. So I would suggest our cranium has become the walls and all that sound is just bouncing around between our own ears. So I would argue that's even more kind of asocial.

DK Yeah—so we need absorptive materials inside our cranium.

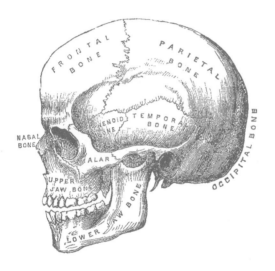

The human skull

ET For all sorts of reasons I think that's true—most people the ideas bounce off and then their own ideas just bounce around in the echo chamber of their skulls. But I digress.

DK Well not really because "echo chamber" seems really crucial, I mean it seems crucial to our moment right now. Which is part of my interest in thinking back to earlier ways that we heard music, or that we heard sound or that we shared sound. Because that's the thing—even if you had no reverberation and you were interested in this direct electrical signal from the stage, you still had this collective experience of it.

Ace Tone EC-20 Echo Chamber tape echo machine, 1975

ET No I think you're right. And I think I would want to emphasize that it takes a lot of technology to isolate each of us in our respective echo chambers—and I think that's not inherent. You know I always say sound wants to mix. For a lot of decades in the 20th century people struggled to control and contain sound. And now we really in many ways have that ability. But why we want to do so—I think we need to step back and question that again.

⌐¬
∟_

In the early 1990s, I had the great good fortune to meet the composer and philosopher John Cage. He received me in the apartment he shared with his partner Merce Cunningham. An apartment on the top floor of an old industrial building, facing one of the busiest avenues in Manhattan.

When I arrived, he was sitting at a table by the window, composing music on paper. And the windows were open.

The roar of the city was familiar from my own childhood apartment. But Cage and Cunningham lived in a building that didn't reach as high in the air. The individual noises from the street below—the buses, the honking, the shouting—were much closer, more distinct, and much much louder.

But John Cage said he never closed the windows. Why would he? There was so much to listen to all the time.

John Cage in his apartment in New York, June 1979

3

[SFX: FaceTime ringtone]

DK Let's see, does that work?

Nancy Harrow Krukowski Yes, yeah, right?
Do you hear me?

DK Yeah do you hear me?

NHK Yes absolutely — it sounds even better
than our phone, yeah.

[BED: Nancy Harrow, "Lover Come Back to Me"]

My mother is a jazz singer, Nancy Harrow. She
was pregnant with me when she posed for the
cover of her second album, for Atlantic Records in
1963. The photo is carefully cropped, so no one
could tell.

I'm sure my mother sang to me as much as talked, when I was a baby. One of my earliest memories is her singing a Nat Adderley tune while keeping time on the soles of my feet.

Nancy Harrow, *You Never Know*, LP, Atlantic, 1963

NHK (singing) Now I reckon that oughta get it, cause I'm working... Working!

DK Right.

NHK It's just that I liked it, it was very rhythmic and when I would be changing you or something that's what I would do.

DK So you would play it on my feet while you were changing me? Do you think my memory goes back that far?

NHK Of course! I think it does.

No wonder I became a drummer. But your mother doesn't have to be a jazz singer, to have memories of being sung to as a baby. Everyone sings to babies. Melody is part of the way we learn language.

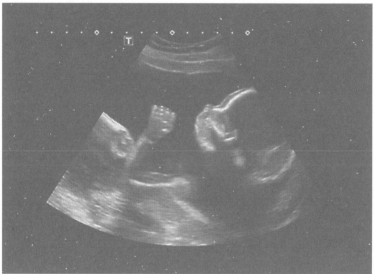

Ultrasound image of a fetus in the womb

Think about it: before we know words, we know the sound of a voice speaking them. Or rather, singing them. Because what is language without words? It's the *sound* of a voice. A musical sound.

NHK I think you know before you could speak you could still understand what I was saying... The child, I've heard, recognizes the sound of his mother's voice because he was hearing it in the womb. That's what people say.

DK So does that mean I heard you do your overdubs on your Atlantic album…?

NHK No but what you really must have heard was when I was in Baltimore and I was playing Ray Charles all day long.

DK Wait wait wait what were you doing in Baltimore singing Ray Charles all day long…?

We understand the voices of those we love so well—even better than we understand their words sometimes.

And when we share our voices—whether through song, or speech—aren't we always saying more than our words? We're sharing sounds so basic to our understanding that they precede language.

But the way we often hear voices today, even of those we're closest to, is mediated by technology. Technologies that extend the reach of our voice—across space, across time. But does *all* of our voice make it through that transmission? Not just the part with the words. But the part that came before that. The part that expresses more than we can say.

Because if it doesn't, are we really saying all that we mean?

Love

Love songs are universal. And there's a universal fascination with those who sing them.

> **Frank Sinatra** I read so many varied reports written by MDs, by psychiatrists, by psychologists, and it ranged from the fact that they were rooting for the boy next door, all the way up to a sex symbol, we might say.

Frank Sinatra, *No One Cares*, LP label, Capitol, 1959

That's Frank Sinatra, speaking in 1965. He's trying to explain why people feel close to him, even if he doesn't know them at all.

> **FS** But I really have no specific answer. I just think that the crooner—or a troubadour, let's put it that way—I think has been a symbol of entertainment that people have been attracted to for many, many years. I think they've admired anybody that can get up and sing a love song.

[BED: Frank Sinatra, "None But the Lonely Heart"]

What is it about his voice that draws us in. Or what is it about the way he *recorded* that voice?

> **FS** I love recording. I think making records is the great fun of all time.

Sinatra not only loved recording, he was a master at it. Singers everywhere still admire his microphone technique. He perfected an address *to* the microphone, carefully controlling his breath and diction so that he could move very close to it, without the disruption of unwanted sounds like the pops and hisses of consonants. So close, it can feel like he's singing right in our ear.

[BED: Frank Sinatra, "When No One Cares"]

FS If any popping of P's let's stop, because there were too many P's popping in past dates we did.

At other times, he backed away from the mic, and let us feel the power of his voice. That constant adjustment of his distance to the microphone makes full use of what audio engineers call "proximity effect."

[SFX: close to mic]
All microphones, if you speak closely to them, exaggerate the bassier, chestier tones in our voices.

And if you back away from them, they highlight the brassier, clearer tones.

Neumann U67 microphone, 1960

Sinatra's control of proximity effect was a part of his communication of *feeling*, in addition to the words of a song.

Not that he wasn't very particular about those words.

> **FS** The biggest complaint I have about a lot of the kids who sing today, is that I can't understand what they're saying. If I could only understand some of the words, I might be more interested in what they're doing. But I can't—there's no enunciation, and there's no clarity of diction.

But it isn't *just* the words of the songs that come through in Sinatra's recordings—after all, many singers of his era could enunciate. There's something else again, about the way Sinatra communicated with his voice.

Something some of those kids he complained about could actually do pretty well, too.

[BED: live Beatles 1964]

When the Beatles swept into the USA for the first time, in 1964—the year right before that Sinatra interview—many older listeners complained that they couldn't make out any of their lyrics.

The Beatles play in Washington, DC, on February 11, 1964

But the *feeling* of what they were singing was clear—clear to millions.

Rock and roll lyrics are so often misheard, there's a whole genre of YouTube videos devoted to spelling out the wrong words to hit songs.

[POST: Jefferson Airplane, live "Somebody to Love"]

In that way, Sinatra was right—if rock and roll depended on enunciation, and clarity of diction, it would be a nonstarter.

[BED: Rolling Stones, live "Get Off of My Cloud"]

John Pasche, Rolling Stones
"Tongue & Lips," 1969

But what Sinatra was missing is that Mick Jagger's *non-verbal* communication over the mic—everything that's not in his words—was not so different than his own. Great rock singers like Jagger may not enunciate, but they do get all the feeling in their voices across.

Feeling that's built on the *sound* of a voice.

Roman Mars I was always attracted to people's voices. My aunt took me to a psychic and recorded the tarot card reading. And it was like an hour long. And I used to play that tape over and over and over again, when I was 14, 15 years old, to go to sleep. I listened to it every single night. So I like this calm, explanatory tone, and I wanted to use that.

If you listen to podcasts—and I know you do—you probably recognize that voice. It's Roman Mars, host of the very popular program about design called *99% Invisible*. I wouldn't be surprised if some people listen to that podcast just to hear Roman's voice—the same way he played that psychic's tape over and over.

DK Did you know you had a great radio voice, like were you told that at some point?

Roman Mars at the PRX Podcast Garage, Allston, MA

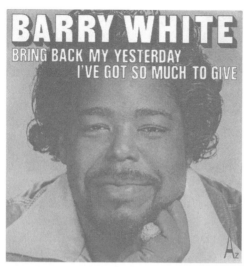

Barry White, "Bring Back My Yesterday / I've Got So Much To Give," 7-inch vinyl, Philips, 1973

RM No, in the beginning I was convinced that I could only be a behind-the-scenes producer, because my voice was so terrible. And so I used to fill in on *Morning Edition* at KALW 91.7 in San Francisco, and because *Morning Edition* you have to get in at 4:30 in the morning, I would come in and barely be awake, have not spoken to anybody, and it would be very much like **this is 91.7**—it would be so low, it would be Barry White-type low, and I just kind of learned how to harness the parts of the calm I could create, the parts that were resonating in my head in ways that I enjoyed.

[BED: Barry White, "I've Got So Much to Give"]

Roman wasn't born with that late-night radio DJ voice. He had to discover it—and develop it, together with a microphone technique. A technique that depends on being very, very close to the mic.

> **RM** In the beginning I would record it in my bedroom at night, when my kids were really little.

[POST: *99% Invisible* show ID]

> **RM** So I had to create something more like a field mic, and just use it really low and really close to my mouth. So that I could kind of work in any untreated room I needed to.

Putting the mic that close to his mouth, Roman exaggerates its proximity effect. Like Sinatra—or Barry White for that matter—leaning into the mic gives us the feeling that he is leaning in close to us, his listeners.

Apple iPhone ringtone volume

[SFX: cell phone ringtone]

> **DK** So we're actually now recording ourselves in two different ways. And I want to reveal that a

little bit—so you're speaking, I take it, on your favorite mic.

RM Yeah.

DK And so we're each here on our favorite mics but then we're also speaking on a cell phone connection, and I wanted to let people hear what that sounds like—the difference —because they know your voice very well over your favorite mic. But a lot of people out there haven't necessarily spoken to you on the telephone.

[SFX: cell phone recording]

RM I don't talk to many people on the phone all that much. I've gotten to the point where people call me—it feels aggressive, you know, like I'm not expecting it.

Touch-tone telephone, 1980s

DK And I assume like me you had analog phones at one point in your life?

RM Yes sure. I mean I remember being 13 or 14 years old and sitting on the floor next to a corded phone in the kitchen for hours on end and never getting tired of it. There were even periods of silence that I remember where nothing was happening or you'd just be watching TV, and you'd still stay on the phone. Now I just don't have any tolerance for that whatsoever. It's one of the rare pieces—a designed object where the quality is so much worse, with weird latency issues and digital artifacts of compressing and decompressing.

I feel like over half my calls end up being interrupted by something—by some kind of sound problem. And it's like, "Well, why commit to anything in depth when there's like a 60 percent chance that it could just end awkwardly right in the middle of your great moment that you're sharing with somebody?"

As Roman points out, the sound of our voices on the phone has gotten worse with the switch to digital. But it's not because the microphones in them have gotten worse—in fact, the miniature mics in our cell phones are much more sensitive than the ones we used in our big old analog phones.

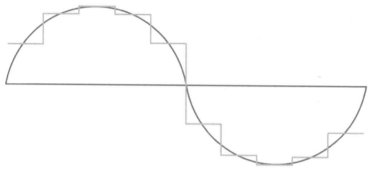

Yet we hear *less* of one another on the phone now. And no matter how close we hold them to our mouths, there's no proximity effect on a cell phone—everyone sounds just as near, or just as far, as everyone else.

The reason is that unlike the old analog phones, cell phones don't transmit the full range of sound picked up by their mics. Instead, they digitally process that sound, compressing it to remove whatever engineers have decided is unnecessary data.

That efficiency makes the remarkable reach of digital communications possible. Just like any other lossy compression—the reduction of recordings into mp3s, photographs into jpegs, film into YouTube videos—the compression of our voices on the cell phone makes for smaller data packets that are easier to share over digital networks.

Engineers know these lossy formats are less than ideal—but they're intended to get the essential information across.

But what is the essential part of our voices, and what isn't? Cell phones are engineered to communicate our words, above all. Everything else is pushed aside—the background noise that indicates where we are. And the myriad small sounds that indicate we're there, too. The sounds of our breathing. The sounds of our *listening*.

That's why we used to be able to hang on the analog phone with one another for hours, sometimes without even talking. We could feel one another's presence.

NHK I was always on the phone.

DK Who'd you talk to the most?

NHK Daddy, and then Liz and Betty.

DK Do you remember all their phone numbers?

NHK I remember all their phone numbers. Ellie's was Havermeyer 9-6448… and Barbara's was Susquehanna 7-4695… I don't remember Daddy's phone number—he must have called me all the time.

Nancy Harrow, *You Never Know*, LP label, Atlantic, 1963

[BED: Nancy Harrow, "Tain't Nobody's Bizness If I Do"]

NHK I dialed the wrong number, I dialed Ellie's number with Liz's extension. And I got this man on the phone who started to flirt with me. I hung up, because it was frightening. And I told the story to Liz, and she kept calling this guy up and gave him her phone number and everything. And she had a long flirtation on the telephone with some strange man, who must have lived in Queens somewhere.

DK She called him deliberately?

NHK Yes deliberately cause that's what she was like, yeah.

DK But you could flirt on the phone, do you think people can still flirt on the phone? That's what I wonder, because it sounds so terrible.

NHK I don't think you can. It doesn't sound the same; it's not as intimate.

DK What is it about that?

NHK Well because you could regulate your voice by the distance you were from it, and the way in which you spoke and whether you made it husky—just like a microphone.

What's lost on the cell phone is the part of our voices that communicates without language—the part we use for flirting. The part that rock and roll leans on. The part that Sinatra was so expert at getting on a record.

Gary Tomlinson is a musicologist who thinks on a grand scale. And he believes that our ability to communicate with the non-verbal parts of our voices goes so deep, it's not only in our memories from early childhood—it's coded into the genetic makeup of our species itself.

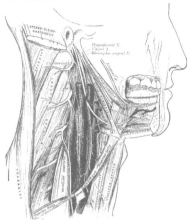

Before we had language, he explains—meaning, hundreds of thousands of years ago—we had utterances. Utterances with musical qualities. And those musical qualities communicated what we needed to survive as a species: toolmaking, social organization. And I would assume love.

> **Gary Tomlinson** Music and language in their deep, deep histories, I see the antecedents of these things falling into place along kind of parallel tracks that overlap one another but are not the same track. And you know, linguistics professors these days they tend to focus on syntax and grammar, and they tend to leave out the musical aspects of language. The tunes of my speech. And if you do that, of course you're going to leave out exactly the overlapping part that is so important.

DK Right. Makes me think also of what's going on technologically for us right now because to encode our language, send it over the internet and make it perceivable at the other end of this great space of distance—the non-verbal qualities of our voices tend to be lost in that coding. Does that relate, do you think, to your sense of the musical aspects of speech?

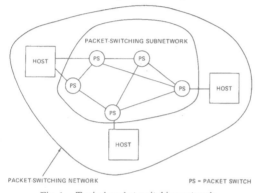

Fig. 1.   Typical packet switching network.

GT There's no question that that's right. There's so much left out—it's so interesting to think about what is absent from what in fact gets reproduced across these technological systems, of which we have so many these days. But the other side is the miraculous way in which human beings fill in what is left out. We—you know so I'm hearing you across FaceTime, there's a lot that is captured of what you're saying and how you say it. There's no doubt other stuff that isn't there. And yet I'm projecting it into it somehow. I'm hearing it. I'm reconstructing you in a way, as you reconstruct me. That's an extraordinary capacity. You know, we are always able to recognize voices almost instantaneously, even though we're getting a small portion of what those voices would give us if we were standing face to face with each other and hearing them. It's quite extraordinary…

DK Right. Oh—

did I lose you?

Gary are you there?
I lost you for a second.

GT I'm here. Hold on a second.

Hey guys, are you online?   *OK.*
Could you stay off for a moment more?

Digital tools make it possible to share our words across great distances. But they fail us in so many ways, as we try to communicate one-to-one. Not only are we often left hanging, speaking into the air at one end and listening to nothing at the other. But even when all is working as it's supposed to, the sound of our voice across digital lines is limited. It's stripped to that minimum we need to recognize a voice, and decode its words.

A speech bubble by Charles Schulz from a 1977 Peanuts comic

But words don't say everything. Gary Tomlinson has a beautiful phrase to express this. He's written that there are "musical absences at the heart of language." We are musical beings, Gary believes, as much as linguistic ones.

And I have to agree. Not only because I happen to play music. And not only because my mother happens to be a singer. But because my mother sang to me.

DK Did your parents sing to you?

NHK My father sang. He would sing acapella in the car always, cause I remember he sang this very haunting song that was about a prisoner in the cell. I mean he told me what it was about because it was in Yiddish. I remember the chorus was "Oy Fagele"—which was the prisoner's looking out and sees a bird flying outside. And it was so mournful, that song, and it always made me feel very mournful.

DK So it was like a blues.

NHK Yes. It was like a blues, it was.

4

[SFX: Van engine starting]

No musician will tell you touring is easy. But I love to travel. One of the benefits of being in a band, for me, is that you get to go places you wouldn't otherwise. Not just the glamorous ones—London, Madrid, Tokyo—those are on everybody's wish list. But if you're on tour, each of those are just one night's stop out of many, many stops.

Ford E-Series passenger van

[SFX: Van door closing]

When you pull into Preston, England; or Huelva, Spain; or Misawa, Japan, it's pretty obvious you're not there to take in the sights. But my partner Naomi and I have had some of our most memorable shows— for good and bad—in those very towns.

A few albums ago, we found ourselves on tour in Europe with a gig in a place we'd never been invited before—for obvious reasons, it had been at war all through the 90s—Belgrade, Serbia. As we drove out

Damon & Naomi on tour in Turkey, 2015

of Italy and into what had once been Yugoslavia, the landscape and language became increasingly less familiar.

And then we hit the border with Serbia. Our cell phones went dead. Our GPS went blank.

We had to resort to old analog touring habits, to find the venue: look for a pay phone. And a coin.

The voice we reached gave us directions: Take the big bridge over the river. On the other side, look for the bombed-out building. Make a left.

The venue, when we found it, was in a student center with a radio station that had been important to the resistance under the previous government. What could we bring to this place, still so barricaded, still so remote from the rest of Europe? There were very few people on the street, and hardly any open shops—there wasn't much food visible anywhere. There were certainly no record stores.

[BED: Damon & Naomi, live "Eye of the Storm"]

Yet when we started to play that night, the audience knew our songs. It was all I could do not to cry on stage.

Our music had made it through all those barriers—barriers that made it impossible for records to get there. Barriers that made it hard for musicians to get there.

Damon & Naomi performing in Tokyo, 2009

Music is immaterial. When we play it, it flows through the air. And when we record it— now, in the digital world—it flows online.

That freedom for our recorded music has compromised our ability to make a living from it.

But I don't remember what we were *paid* in Belgrade. There was so much else being exchanged that day, between us and the audience.

Episode 4

Money

Most musicians I know are paid much too little, or much too much. Maybe it's because no one is sure what music is really worth.

> **Dave Grohl** The thing about music is that music should be available to anybody that wants to hear it. I think that there should be no such thing as a price tag on music.

That's Dave Grohl, the drummer in Nirvana, on a cable TV chat show in 2001. He knows full well what kind of money can be made from music. But that doesn't mean he thinks it should have a price.

Dave Grohl on *Dennis Miller Live*, HBO, 2001

> **DG** Ok maybe there's a price tag on the package that you buy. You know you pay $13 for a CD,

you get the artwork, you get the jewel box, you get the friggin sticker, you get the whole nine whatever. But I don't want to turn on my fucking radio and have to put a nickel in it to hear Metallica. I don't want to have to pay to hear music. [applause]

In 1999, a laconic freshman at Northeastern in Boston, Shawn Fanning, figured out a way for people to share digital music they kept on their hard drives directly with one another. He used his internet chat handle, Napster, as the name for the program he wrote—

Napster logo, c. 1999

and it quickly became shorthand for the collapse of the $14 billion music industry.

Shawn Fanning I love music. But you know the technology was probably the most powerful piece for me... This system, what's most interesting about it is you're interacting with peers. You're exchanging information with the person down the street.

Record companies panicked at the idea of their products being exchanged without money involved. Piracy and bootlegging they were used to—those are considered a normal cost of business. But sharing was seen as an existential threat.

Not by musicians. Apart from some very vocal, and litigious exceptions—like Lars Ulrich of Metallica, and Dr. Dre of N.W.A.—a lot of musicians saw sharing music online as a natural, inevitable, or exciting technological breakthrough.

> DG I understand where some people come from when they say Napster is taking money away from me.

Here's Dave Grohl again, throwing shade on his fellow drummer Lars:

> DG But you know what? When it's someone that's sold 50 million records and they got 50 million fucking dollars and they're bitching about pennies? Fuck you, man. [applause]

And here's Chuck D of Public Enemy, framing the issue in more historical terms:

[BED: Public Enemy, "Fight the Power"]

> Chuck D Cause the industry has, over the last 50–60 years, has been accountant and lawyer driven, and it hasn't been about the artistry. I look at Napster as power going back to the people.

Public Enemy logo, Chuck D, 1986

I also look at this as being a situation where, for the longest period of time, the industry had control of technology. And therefore the people were subservient to that technology. At whatever price range that the people would have to pay for.

You won't find Chuck D today selling cheap headphones for millions in profit.

But Dr. Dre's subsequent move into music technology—his headphone company Beats is now a multi-billion dollar part of Apple itself—proves Chuck D's point perfectly. The music industry—as an industry—was always focused on the sale of technology. And not necessarily music.

After all, it's hard to sell something immaterial.

The tenuous relationship between music and money was forged at the turn of the previous century.

In the Library of Congress there's a room not many people visit anymore. It has rows and rows of cabinets with drawers crammed with index cards. Each one lists a song. And its copyright holder.

DK Well let's see what I can find, I'm going to look up under "swing"—swing, swingin', swinging—there's just enough room left in the drawer to just nudge apart the cards. It's like a record store that's jammed everything into the bins...

This *was* the music business, before recording was invented. Sheet music publishers controlled the hits of their day. Every parlor that boasted a piano needed sheet music to play on it.

And then in 1898, a hundred years before Napster, a digital invention disrupted this industry. The Pianola, or player piano.

The player piano didn't need any sheet music. Instead, it used a roll of paper with holes punched in it—each hole is a digital instruction: on or off, one or zero. As the roll turned, the holes triggered keys on a piano, and music happened.

The music business panicked—and called their lawyers. Aren't these player piano rolls violating our copyrights, by playing tunes whose titles we wrote down on index cards in the Library of Congress?

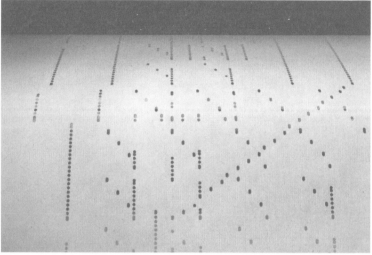

Music roll running through a Gotha Steck Model 15 Pianola Piano, 1913

The case went all the way to the Supreme Court. And in 1908, the decision came down in favor of the new digital technology. Music publishers, the court reasoned, owned their copies of written music—that's what *copyright* means—but not the music itself. They couldn't, said the court, because music is not tangible. Sounds are in the air—they're immaterial—and therefore can't be property.

But stamped into the pressing of each and every record are other copyrights: the label, the logo... All those printed bits the record companies *could* own.

Copyright symbol, introduced in the United States Copyright Act of 1909

A record is like a sandwich of tangible, saleable materials—vinyl, paper, ink—with an intangible, unownable center. The music.

What Napster did that was so damaging to the industry, was to separate music—which doesn't have a price attached to it—from the technology that delivers it, which does.

But when Napster turned its servers off, by court order, the files didn't stop flowing.

[BED: Jace Clayton, *Gold Teeth Thief*]

Jace Clayton I made this mix called *Gold Teeth Thief* in 2001 and mailed the CD—I was in Spain at the time—mailed the CD to my friend in Cambridge and I was like, can you like make mp3s and put this online? And he did. And then you know six months later I was getting international invitations to DJ. I was like—this is—it kind of blew my mind…

That's Jace Clayton, who performs as DJ Rupture. Jace is also a writer who has analyzed the flow of digital music—and music tools—around the globe.

DJ Rupture - Gold Teeth Thief.mp3

JC I was using vinyl to make that mix. But then that thing traveled because it was, you know, this slippery mp3 file. So there was like this moment of global listening that really kind of kickstarted my career as an international DJ.

Jace has investigated some fascinating ways that digital music tools developed in the first world have made their way to the third—like AutoTune in North Africa, where a technology intended to polish notes in Western pop has been put to entirely unforeseen uses.

[BED: Hafida, "Youchkad Zin"]

JC One of the ways I researched this was spending time in studios in Morocco, and like nobody cares about the software interface, they turn one knob to zero and one knob to 100 and then they just go. Because what's desired is just the weirdness of it.

Hafida, *Ayan Dar illa Zine*, CD, Fassiphone, 2007

DK That is such a wild story. But it's also just really exciting, the idea that something in California will affect Berber wedding music.

JC Yeah—and you know what it is, it's like we're not talking about the aspect of Berber folk culture that goes worldwide and changes the way guys in Silicon Valley are programming databases. Unfortunately. You know, it's still this kind of a center-to-periphery flow. But there is—I'm very interested in the ways in which all sorts of digital peripheries can sort of like reinscribe,
            retranslate,
                        mistranslate and
pull   apart
some of these—some of these structures that are given.

That tantalizing possibility of change coming back from the periphery to the center is not just a technological dream—it's very much a political one. If we keep the internet truly free, there's no telling the progress we might make through its open lines of communication.

[BED: Downtown Boys, "Heroes"]

Aaron Swartz Big stories need human stakes. But that's kind of the point—we won this fight because everyone made themselves the hero of their own story. Everyone took it as their job to save this crucial freedom. They threw themselves into it. They did whatever they could think of to do. They didn't stop to ask anyone for permission.

Aaron Swartz, 2009

That's the voice of the late internet activist Aaron Swartz, sampled by the Providence punk band Downtown Boys for a track called, "Heroes." He's talking about a grass-roots campaign he waged to defeat a bill in Congress that would have compromised net neutrality—the principle that all data on the net should be treated equally by the platform. From the periphery, to the center.

[BED: Downtown Boys, "Somos Chulas"]

    Net neutrality may sound like a dry topic for a rock band. But it's one of the rallying cries Downtown Boys have placed at the center of their very dynamic music.

    Downtown Boys were recently signed to Sub Pop, the same label Dave Grohl started on with Nirvana, and the same label Naomi and I recorded for throughout the 90s. I wanted to find out how the music business looks to them, coming up as they have in this post-Napster, digital music world.

I caught up with the band backstage before a show in Boston.

Downtown Boys, 2017

Downtown Boys We sold fifteen records last night, we have a tally. Yes I mean we're still all broke as of now.

Some things don't change. But this is a band with way more political awareness—and vocabulary—than I remember anyone having on our scene.

Victoria Ruiz I mean like the cost of living is just higher...

Here's singer Victoria Ruiz, talking about *all* the costs of living. Not just the monetary ones:

VR And then we also pay more for like the emotional labor that I think I take on as a woman of color fronting the band, and I think Mary and Norlan take on as people of color in the band. Like right now I think because of the internet you don't see that, you don't see what's behind. You're in the front but then you're also in the back like pushing it, and no one sees you when you're in the back. And that's where 99% of the work happens, all the time.

Joey DeFrancesco And right now the monetary reward for that is just like ever ever ever decreasing.

And this is guitarist Joey DeFrancesco:

JD There's money being made—it's just the distribution of it is worse than it was twenty years ago, which you know the internet is supposed to equalize but hasn't. And you know I think we have to do some kind of organizing as musicians and like take seriously that we are laborers and can make demands. Cause like a lot of capitalism, the whole industry is designed to disguise the labor. So like I used to work in a hotel and they would you know really try to disguise like the housekeepers' labor—like the housekeepers come out when most guests are out of the hotel, and they're supposed to get to the room. The labor is supposed to be completely invisible behind it. But with like a record, it's just supposed to emerge and you have a big release day and like everything behind that— writing the songs and making this thing and like doing all this logistical work to make this happen—just gets like completely erased.

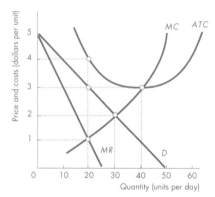

Profit maximization diagram

VR Yeah like I studied economics or whatever and like they teach you this kind of classic market economics graph, and it's like the way to maximize your profits is to minimize your costs. And we're a band that we will never be able to minimize our costs—because we are poor, we're people of color, we're queer, we're coming from all these histories, all these backgrounds. The costs are really high. And so the profits will never be like this maximized thing and that's not why we do it. It's way more of like a reciprocal... The best parts of the band are when we can reciprocate with the audience and break down that divide between the audience and the crowd and us.

2014 , the tobar ,Baltimore

[BED: Downtown Boys, live "Ceremony"]

The way Downtown Boys talk about their material
situation takes account of money—but it's not in terms
of copyright, royalties and piracy.

Instead, it's in terms of labor. They see the struggles they face of a piece with the political struggles
of this 21-century moment. They don't see their life as
musicians set apart from those issues—on the contrary, they see it enmeshed in them.

When I first started touring, with Galaxie 500 in the
late 80s, it felt like running away to the circus. I jumped
in the van and we drove away—away from the jobs
many of our friends and classmates were heading into.
And away from the ties those jobs clearly had to the
larger economic and political world.

Money   87

The music scene we were part of—a more or less underground, indie one—seemed like a different place. One that operated largely by sharing—trading information, trading zines, trading records. Promoting one another's shows. Sleeping on each other's floors.

Galaxie 500 backstage at La Dolce Vita, Lausanne, Thanksgiving Day, 1989

Eventually, the world caught up with Galaxie 500—in the form of major label record companies—and like so many bands, we split up over some of the same issues I thought we had left behind: chief among them, money.

When Naomi and I started touring again as a duo, in the mid-90s, we vowed to never again let the material world distort what we love about music.

And it's been pretty easy to keep that promise—because in the digital era, musicians are very clearly operating in an immaterial world.

Our digital recordings have no set value—they earn us pennies at best. But just like we have ever since we first jumped in a van, they love to travel.

That's how our music made it to Belgrade, long before we could. But it's not just places at the periphery that can change from free communication online. It's the center, too.

Like here in Boston, where Downtown Boys have just launched into a cover of "Ceremony," by Joy Division. A song I know very well, cause Galaxie 500 used to play it too.

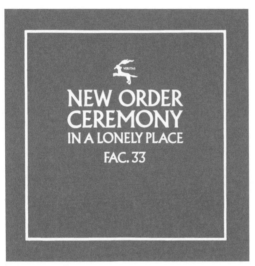

Joy Division, "Ceremony," released as New Order's debut single, 12-inch vinyl, Factory, 1981

"So what is it? What is *your* forever?" Victoria riffs toward the end, leaving the original lyrics behind.

"How do you hold that contradiction," she asks. "Being completely finite, completely limited, but believing in forever—believing in something bigger than yourself?"

The terms of exchange for music today aren't about money, but that doesn't mean they aren't very, very real. Maybe realer than they were, when we were all busy pretending that music could be simply bought and sold.

5

Stereo Jack's, Cambridge, Massachusetts

[AMBI: DK at Stereo Jack's]

I still go to record stores, especially used ones. It's not that I'm a vinyl snob—at least, I don't think I am? Anyway even if I am, these days you can find any particular record you might be looking for much more easily online, at Discogs or eBay or even Amazon.

But I go to record stores for what I *don't know* I'm looking for. I go to stumble on something, by chance.

And even if I don't find anything to buy, I always pick up some knowledge from whoever else is there. Record collectors and record store clerks can have very out-there interests. And strong opinions.

They also have a lot—a *lot*—of information. If you've been sorting through records for years on end, you not only know just about every record I might ask about, you know the label it was on, the year it was released, the likelihood of finding it, and probably a couple good anecdotes about the players on it as well.

I go to record stores because there's actually no other good way to find most of that information, online or off.

ERIK SATIE, "Parade," performed by Georges Auric and Francis Poulenc, Éditions de La Boîte à Musique, No. 16, 12-inch 78rpm label, 1937

[BED: Erik Satie, "Parade" piano duet, performed by Georges Auric and Francis Poulenc]

I also go for the same reason André Breton, the Surrealist artist, went to the flea market. Here's what he wrote in 1928:

"I go there often, searching for objects that can be found nowhere else: old-fashioned, broken, useless, almost incomprehensible, even perverse."

Breton was describing the crate digging of his day. Because crate digging is not just shopping. It's sorting through bins of whatever current mainstream culture has no use for. Even yesterday's *hits* end up in the dollar bin, eventually.

Zager & Evans, "In the Year 2525," 7-inch vinyl, RCA, 1969

[POST: Zager & Evans, "In the Year 2525"]

And there's something very powerful in recuperating what power has rejected. André Breton considered himself not just an artist, but a revolutionary. He was interested in overturning society's expectations in every way—elevating the ignored, to bring it out of the shadows.

How else, he reasoned, can we surprise one another? And ourselves?

The marginal—the rejected—the repressed— is whatever the powerful have decided is of no use at the moment. But might it not be a key to alternate approaches—to art, to society—to power itself?

Episode 5

Power

The internet gives us access to so much information, it creates a power vacuum: what will order all this data for us? What will bring a sense of hierarchy and priority to it?

> **Jimmy Johnson** This is the shipping department, these are orders right over here they're being prepared, and there's tape machines over here.

That's Jimmy Johnson, who runs the independent music distributor Forced Exposure.

Forced Exposure's warehouse, Arlington, MA

Wave-like sand ripples formed by wind or water

JJ If you think about it, digital music is like grains of sand or something at the beach—like, it just goes as far as you can see. And there's like no reason to think that any one of those grains is any better than any of the others. You would never build a shelf to store your grains of sand. And you wouldn't go online to a service and just say, "I'm just going to listen to all the music that's there." Like I definitely remember going to Tower and saying, "I want to hear everything that's in here." But with digital music it's like, well, if it just goes past the sunset, like I don't care about any of it.

Jimmy Johnson is not daunted by a large quantity of music—when Tower was the biggest record store in the world, he wanted to hear everything in it.

Today, Forced Exposure carries 50,000 titles, with 100 new ones arriving each week—and Jimmy listens to them all.

> **JJ** And this is my office right here. These are basically all the recent new releases on the floor of my office. Insurance agents come through and say please move these.

Jimmy Johnson's office at Forced Exposure

What's more, he has his whole staff hear them, too. Jimmy's the DJ in the office, and makes sure to play all the new arrivals.

> **JJ** I feel like it's really instructional for our employees to look at every record, like, this is someone's *work*. You can't just pass it on along and say, "Well, I gave this two minutes of my time and I don't care."

DK That's distribution as responsibility.

JJ Distribution as responsibility, yeah—and maybe not everyone feels the same way about that. But it's a challenge.

The Forced Exposure staff not only listen to every release they stock, they write about them: 75,000 words a week. Each month they produce a printed catalogue that describes every record in detail.

JJ So I've never admitted that we've become a distributor. You know it's all a faux cover-up really—we're a deceased magazine.

DK With a very big warehouse. And a basketball court.

JJ And a basketball court.

Basketball is another of Jimmy's passions. Fortunately, it's no problem fitting a court into a warehouse that's big enough for all those records and CDs.

Jimmy Johnson shooting hoops in the Forced Exposure warehouse

Forced Exposure started as a zine in the 80s—Jimmy and his fellow writer Byron Coley listened to all the music they could get their hands on, and told the rest of us what they thought was worth hunting down. And back before the internet, it really was a hunt.

JJ Especially foreign territories, like I remember getting into Finnish hardcore in the 80s—but you had to like write a letter with a self-addressed stamped envelope. So you're talking months or even years just to track someone down. It wasn't like you could just look it up—it was pre-internet. And in that time waiting you kind of build up an insatiable desire to find out.

All that finding out slowly transformed Jimmy's magazine into a distributor. A distributor I and many others have followed closely, to learn about music from all over the world that we could have never found otherwise.

It was Forced Exposure that first brought records by the Tokyo band Ghost into the US— records that then inspired me and my partner Naomi to seek them out in person, and eventually collaborate on an album together.

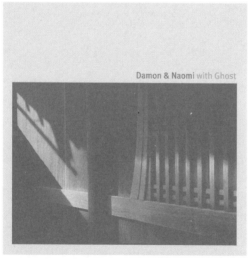

[BED: Damon & Naomi with Ghost,
"The Great Wall"]

Damon & Naomi, *Damon & Naomi with Ghost*, CD, Sub Pop, 2000

I can't imagine that back then, if we'd been able to somehow tap directly into all the music happening in the world, that we would have found Ghost. It would have been like sifting grains of sand, as Jimmy says. But stumbling on them in the Forced Exposure catalogue—a large but limited set of releases, each one listened to and chosen by Jimmy Johnson—feels like it was inevitable. We were destined to connect.

Fifteen minutes down the road from Forced Exposure's warehouse is another business committed to music recommendation. But the similarities end there.

> **Paul Lamere** I think my ultimate goal is this magic music player that automatically knows what you want to listen to.

The Echo Nest logo

That's Paul Lamere, who designs recommendation programs for The Echo Nest, a division of Spotify. It's his job to figure out how to get you the right grains of sand.

PL So instead of having to scroll through millions of songs to figure out what you want to play, you just hit a play button and it plays the right song for you based on your context your mood where you are or what you're doing, without you having to think about it.

Spotify iOS app icon

Digital music companies like Spotify, Apple Music, and Pandora want to work with the entire universe of available music. Forced Exposure's 50,000 titles are a drop in the ocean for them. Which means no one at those companies—no one—is listening to everything. It's impossible. It's not a human task, on a human scale.

Which means it's done by computers, using algorithms.

PL Probably around 2005 the way people thought about music recommendations was very similar to the way they thought about movie recommendations—if you liked *Jurassic Park*, you might like *The Matrix*. Well turns out when you do that with music recommendations

Apple Music iOS app icon

you get all sorts of funny problems. First of all you get this real popularity bias especially around music releases, so it may be that Michael Jackson and Sting both released albums the same week and even though they're very different types of music, because they're released the same week people may be listening to them at the same time. So you maybe start to get recommendations of "if you like Michael Jackson, you like Sting"—which is actually not a very good recommendation.

PL So sort of the realization was, social recommendation is not enough for music. So one of the technologies that we started to look at was all-around acoustic similarity. That Michael Jackson and that Sting song—they actually don't sound the same. Well, turns out that was a challenge as well. They had all sorts of trouble where a harpsichord actually has a waveform that's quite a bit similar to a distorted guitar. So you put on some heavy metal and you start to get Bach canons recommended. That's not a good thing.

Glenn McDonald/Spotify, "Every Noise At Once" genre map of over 1,100 types of music at everynoise.com

PL So the next wave of acoustic recommendations sort of gave way to the third wave which was cultural recommendations—and this is trying to understand music based on what people are saying about the music. So online there's people writing thousands and thousands of album reviews for popular albums. So the idea is to take all of this text, do data mining on the text and get an idea of what the music sounds like based on that. So again this technology worked really well, but again had some flaws. For a brand-new artist where you don't have those 10,000 reviews, you don't have a good way of doing the recommendations.

PL So the next phase is, well, let's combine all of these approaches—so the hybrid approach of using the cultural, the acoustic, and the social data. That's sort of where we are now.

Despite the problems in each of these approaches, where we are now, in Paul Lamere and his colleagues' recommendation algorithms, has become a pretty amazing place. I use Spotify's "Discover" feature, and its predictions of my musical tastes are so accurate it can be unnerving. How did they know I would like this record that I've never heard of, never seen in all my record browsing, and yet sounds so much like other records in my collection that it somehow feels like I've heard it before?

Maybe because it *is* like something I've heard before. It's not *surprising*.

Surprise is not the same as "discover." To a huge digital corporation, eager to engage every one of us and as much of our time as possible with their product, surprise is not really a helpful thing. Does Google want to "surprise" us when we use it to search for something?

Does Facebook want to "surprise" us when we look for our friends?

Not at all—they want us to find what we're already comfortable with. Google is engineered to provide us with direct answers to our questions—not Surrealist replies that turn our questions upside down, or challenge why we were asking them in the first place. Facebook wants to connect us to the people we already know, or at least know of—not those we've never encountered before.

And music recommendation services like Spotify want to give us the music we probably like. At least enough to let it keep on playing.

Ghost, *Ghost*, CD, P.S.F. Records, 1990

[BED: Ghost, "Sun is Tangging"]

Which is not something that sounds like nothing we've ever heard before. That could be the best thing we've ever heard—that's how I felt when I got the first Ghost albums from Forced Exposure—or it could be the worst, which we would click away from, before any ad tracker had the chance to tally our attention.

Keeping our attention—or at least, keeping us engaged inside their program—is at the moment the goal of some of the most powerful corporations in the world.

Online, it's becoming more and more difficult to escape the influence of those corporations, and their algorithms that shape the subset of information we each see. They are replacing the freedom and chaos of the internet at large, with the control and predictability of their programs.

But subverting that system is easy *offline*. It can be as easy as walking into a bookstore.

City Lights, San Francisco

One of the best bookstores anywhere is City Lights, in San Francisco. Not because it's the biggest, like Amazon claims to be—it's in the same small building where poet Lawrence Ferlinghetti first opened it in 1953. It's fantastic because the stock is so carefully chosen. And the staff is so knowledgeable.

Up a partially hidden staircase at the back are the offices for City Lights, the publisher. Elaine Katzenberger, Executive Director, knows both sides of this business—publishing and retail—and has been managing its survival into the digital age.

City Lights Executive Director Elaine Katzenberger

Elaine Katzenberger It used to be the rhetoric was kind of like, oh well you know it'll be like LPs. You know I was like no no no there's nothing artisanal about this, or precious, we're not just going to become a rarified place where it's a museum to how books used to be. Like this is alive!

Elaine and I talked about what the experience of a store like City Lights means, in this era dominated by digital corporations and devices.

> **EK** We as humans we have our desires and we have our needs. And so if I want to walk over there the best thing would be if I could just walk straight through you and everybody would part. And then I would just get there. So that may be a human desire, but most of the time the world doesn't accommodate human desire. And so you have to accommodate yourself to things. But what digital means is parting the waters all the time and all you get is your preconceived experience.

Finding what you already know you want is what has made Amazon so useful to book buyers in the digital era. You can reach out and pluck exactly what you're looking for from its virtual shelves, which are endlessly long. No physical bookstore can compete with that.

> **EK** So it's kind of like well OK—when all you need is one thing and you really don't need anything else, then fine. Go ahead—but realize that what you're missing is everything else, and really realize that. So have the other experience

to remind yourself like it is a big world out there and it's a physical world out there. And it's a much richer world than just what you think you need. So no, a place like this shouldn't be lost. But it shouldn't be patronized in that way, like oh I have to give them some money so they won't be lost. It's more like I don't want this experience to go away. You know—I need this experience for my soul.

Just as Jimmy Johnson feels a responsibility to the records he distributes and the work they represent, Elaine Katzenberger sees books as much, much more than units. And shopping for books as something other than an exercise in efficiency.

But City Lights is not an island, nor does it want to be. No one involved in communication wants to be.

Amazon Fulfilment Centre, Rugeley, United Kingdom

EK Well as a publisher you know we have to sell our books on Amazon or our authors would never publish with us. And it is a very galling experience because they're a terrible company to work with. I mean, "work with." That's kind of even just—that's not even true. So yeah you can't choose not to do business with them, and you can't choose the terms of doing business with them.

DK That's really crucial. You don't choose the terms—it's not...

DK and EK together ...it's not a negotiation.

DK And I think that everybody can sense that, you don't have to be producing books or records to understand that because you have those terms of agreement that we all click through and just say yes yes yes, just stop showing me these screens—because it's not a negotiation right. But that is technically a contract, it's just that we have no say—we have just opt in or opt out. And if you opt out, you really opt out.

Before you can complete your registration you must accept the terms of service.

I Agree

View Terms

Cancel

If opting out isn't an option for you—and you're listening to this podcast, so I know it's not—then you're probably engaging with giant digital corporations all the time. That engagement may seem the same as searching out information in physical space, just more efficient. But it's fundamentally different.

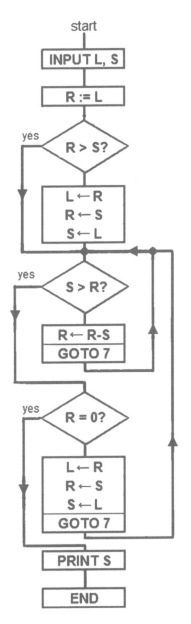

When you go into a bookstore, or record store, or library—any physical space devoted to information—you enter another world. A world that may be confusing at first, even overwhelming, but one that you learn to navigate.

You adapt to *it.*

But when you open a browser— it's an irony that's the word, isn't it?— that relationship to information is reversed.

It conforms to *you.*

At Spotify, the dream is to provide you with music without your participation— the *algorithm* will know what you want. At Google,

Apple, Facebook, the strategy is the same. They all mine data collected about you, to shape the world of information that you find inside their programs.

How people bring your info to apps they use

People on Facebook who can see your info can bring it with them when they use apps. This makes their experience better and more social. Use the settings below to control the categories of information that people can bring with them when they use apps, games and websites.

- [ ] Bio
- [ ] Birthday
- [ ] Family and relationships
- [ ] Interested in
- [ ] Religious and political views
- [ ] My website
- [ ] If I'm online
- [ ] My status updates
- [ ] My photos

- [ ] My videos
- [ ] My links
- [ ] My notes
- [ ] Hometown
- [ ] Current city
- [ ] Education and work
- [ ] Activities, interests, things I like
- [ ] My app activity

If you don't want apps and websites to access other categories of information (like your friend list, gender or info you've made public), you can turn off all Platform apps. But remember, you will not be able to use any games or apps yourself.

[Save Changes] [Cancel]

Facebook privacy settings, 2012

You find the answers you want to the questions you already know to ask. You find the opinions of those who already agree with yours. You find the news that reinforces those opinions, and the others who are being given that same information.

This may make for an ideal experience, if *all* you want is what you *want*.

But what if you're looking for something else?

And what if that turns out to be something these powerful corporations have no interest in providing.

6

220 Hz, 125 dB

[SFX: Hearing test tone]

[AMBI: DK at doctor's office]

I try and have my hearing tested regularly. I think
of it as dealing with an occupational hazard—
I started my career as a musician behind the drums,
and drums are *loud*. So are amps, which want
to be heard over the drums. And monitors, which
need to be louder than the amps.

It's a vicious circle of volume on stage. As
Deep Purple's Ian Gillan put it, on their live album
Made in Japan, **"Can I have everything louder
than everything else?"**

[BED: Deep Purple, live "The Mule"]

But these days, we're all at risk of hearing
loss regardless of occupation. We're carrying
mini-amps around with us in our smartphones,
and sticking tiny monitors in our ears—it's like
everyone's on stage all day long.

The World Health Organization issued
a report this year warning that 1.1 billion young
people are now at risk of hearing loss from

personal audio devices—and what they called "unsafe listening practices."

But they're shouting in the wind. Not only has digital made audio more portable and more prevalent, it's made it louder. In the first twenty years after the introduction of the CD, the average music recording grew louder by a factor of ten.

Commuters with earphones on the New York subway

[BED: Metallica, "That Was Just Your Life"]

These "loudness wars," as they've come to be known, have settled into a kind of détente now, and somewhat leveled off. But no one has gone back to the levels we used in analog days.

The reason is that digital allows the boosting of signal, without the boosting of noise.

Signal and noise are always joined together in the analog world. Think of an LP, with its surface noise. If you turn up the volume, you turn up the noise too.

Now consider a digital signal, like this podcast. There's no noise on it from the medium — no surface noise like an LP, no tape hiss like a cassette, no static like on the radio. So if you turn up the volume, you get... more volume.

In the digital audio world, everything *can* be louder than everything else. There's no noise to restrain signal.

Metallica, *Death Magnetic*,
CD label, Warner Bros., 2008

But if everything is loud, is anything really audible? There's more to noise than the scratch of an old LP.

Signal & Noise

I learned to think about sound in terms of signal and noise in the recording studio. But it turns out doctors use the same terminology.

> **Alicia Quesnel** You know if you just think about what noise means I think that's probably the signal that you're not interested in. And so a lot of what we do as ear surgeons is to try to amplify the signal, and decrease the noise.

That's Alicia Quesnel of Harvard Medical School, otologist and surgeon at the Massachusetts Eye and Ear Infirmary. And my hearing doctor.

> **AQ** But the—you know the ear is just a beautifully designed instrument naturally where it relies on the richness that you get from having 30,000 neurons that people are born with, and which allows your brain and in combination with your actual peripheral ear to be able to sort out what the signal is and what's important to listen to, and to be able to hear that above the noise.

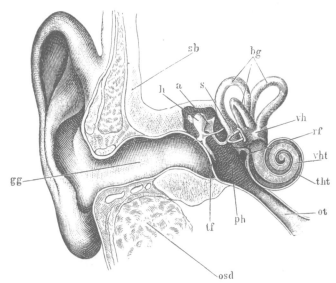

To Dr. Quesnel, signal is whatever sound we are trying to pay attention to. That's a constantly shifting target—picture yourself in a crowded restaurant. Your dinner partner is saying something across the table that you want to hear: that's signal, and everything else in the restaurant is noise. But then someone at another table says something of interest to you—as you eavesdrop, that voice becomes signal, and whatever is being said at your table is now the noise.

We're very skilled at shifting our attention from signal to signal, focusing and refocusing on different sounds in the environment, and shutting out others. As Dr. Quesnel says, we have a rich field of sound to work with—30,000 neurons' worth, at least to start.

And all of that is noise, until we decide what is signal.

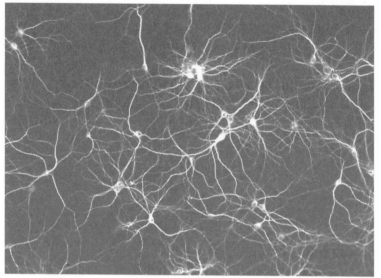

Sensory neurons respond to one particular type of stimulus such as touch, sound, or light

AQ You know the richness of noise includes everything that you're hearing—it's everything that you're hearing, the background noise that you may not be paying attention to. Obviously your brain tries to pay attention to the signal, but there's a lot of other sounds that are going on that are constantly being recorded essentially or sensed by your ears.

DK So everything is in there in the noise, all the sound is in the noise.

AQ Right, right.

What happens with hearing loss is we lose some of the richness of that noise, and with it our ability to distinguish differences in it, and pull whatever signal we're interested in away from the rest. That crowded restaurant—as anyone who has dined with someone older knows well—becomes a mass of indistinguishable sounds.

"Can I get you any more deafening loudness?"

Bruce Eric Kaplan, *New Yorker*, 2013

And hearing aids don't help that situation—they just amplify all the same sounds, making them louder but no more distinguishable from one another.

It's the same problem as the loudness wars—volume alone can't help us find the signal in the noise.

[BED: Shellac, "Il Porno Star"]

In a recording studio, microphones open up a rich field of sound, just as Dr. Quesnel describes for our ears.

But instead of our brain, it's then up to an audio engineer to decide what in that is signal, and what is noise—maximizing one, and minimizing the other.

That decision might not have anything to do with what we typically call "noisy." Think of a distorted electric guitar playing through an amp—the guitarist might be making very deliberate use of noisy sounds.

> **Steve Albini** I deal mostly with punk rockers and people who have cheap equipment, so the noise just comes along, you know.

That's Steve Albini, the meticulous engineer well known for recording bands like the Pixies, Nirvana, PJ Harvey, and many many others, as well as his own noisy music.

Steve Albini at Electrical Audio in Chicago, 2015

But even in punk rock, there are noises an engineer wants to highlight, and ones they don't. Good audio engineers are very attuned to distinctions in what others might hear as simply noise.

SA OK it's an obsession, all right.

Steve runs Electrical Audio, a recording studio he built together with his bandmate and fellow engineer Bob Weston. Together they told me a story about an unwanted noise that proved particularly hard to minimize. We spoke in front of an audience in their hometown of Chicago.

SA We did have a kind of a rage-inducing episode. There was this ratcheting noise: ahh-ahh-ahh.

Guitar amps at Electrical Audio

Bob Weston On a guitar amp.

SA Yeah on guitar amplifiers. We had an electromagnetic interference specialist fly in from Canada.

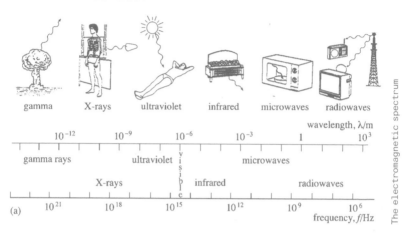

The electromagnetic spectrum

BW It was nuts.

SA One of our engineers, a guy named Greg Norman who's one of our engineers, he built a thing to...

BW ...like a direction rangefinder.

SA Yeah. Well there were two things. My favorite one though is he built a thing that was a Fender guitar pickup on a fishing pole and a battery-powered amplifier, and he walked around the neighborhood with this thing sniffing in the air. And we made an elaborate map that had the studio and then concentric rings showing how loud the buzzing was. And it turned out that there was a photocopier that was in this guy's office—and it wasn't even their main photocopier. It was just the one that he like when he needed to photocopy Snoopy to put on his door or whatever, you know.

And so then we had this routine, like if the noise ever showed up we called the receptionist at the printing company next door and said hey this is the studio next door, could you go ask Bob to shut off his photocopier for us.

[BED: Galaxie 500, "Listen the Snow is Falling"]

The engineers at Electrical Audio needed to trace that unwanted noise, so they could better record the signal they were after. And that's a decision constantly being made in the recording studio. From the placement of mics, through the balance of different sounds in mixing, all the way to the final stage of mastering, audio engineers are choosing what to highlight for listeners as signal and what to minimize as noise.

But one of the reasons I love analog recording is that you can never completely eliminate the noise. Even after fishing for it with a pickup all around the neighborhood, there's going to be something in there more than just the signal you were after. It's in the nature of the medium.

The Beatles, *Sgt. Pepper's Lonely Hearts Club Band*, LP, Parlophone, 1967

Turn up the volume on the paragons of analog recording—like the Beatles' *Sgt. Pepper's*, or the Beach Boys album *Pet Sounds*—and you'll find all kinds of sounds alongside the signal that has been put in the foreground. Those recordings are rich with noise.

The Beach Boys, *Pet Sounds*, LP. Capitol, 1966

[BED: Beach Boys, "Wouldn't It Be Nice"]

Brian Wilson The tracks, the drums, and the way the guitars and pianos were mixed together you couldn't tell—it's just one big sound you know.

That's Brian Wilson, the songwriting and production wunderkind behind the Beach Boys' most elaborate recordings. His studio techniques are admired by musicians and audio engineers of all stripes, because of the remarkable sound textures he was able to weave together. He based that approach on his hero Phil Spector's idea of a "wall of sound."

BW In *Pet Sounds* we utilized what we called the Phil Spector technique—which in the 20th century it's quite a scientific advancement in music and sound. Namely that it's taking instruments and combining them to form them into a new sound. In other words if you took a piano and a guitar and combine them properly soundwise, mix the two sounds together, you come up with a new instrument, or a new sound.

Brian Wilson was interested in the combination of sounds, hearing what happens when they rub together—not in isolating them one from the other. Which means, as meticulous and demanding a producer as he was, he wasn't particularly concerned about noises on his recordings.

In an instrumental break to the *Pet Sounds* track "Here Today," you can even hear a casual conversation taking place in the studio that was accidentally picked up by the mics.

[BED: Beach Boys, "Here Today" original mono mix, conversation starts 1:55ff]

Brian Wilson directing from the control room while recording *Pet Sounds*, 1966

Noise is unavoidable in any live recording, because mics are like our ears—they hear everything. The signal they pick up is always surrounded by noise. Or you might say, everything they pick up is noise— producers and engineers simply direct our attention toward what they want to highlight as signal.

Today, digital recording has a very different way of working with signal and noise.

Take an extreme example of digital music—one made entirely in a computer, without any physical instruments or microphones. You can now make music like that even on your smartphone, using Garage-Band or another program with virtual instruments.

Apple's iOS GarageBand running on an iPhone 4S, 2011

But if you start to combine many virtual instruments together, you may notice you don't get a wall of sound like Phil Spector's or Brian Wilson's. Instead, you get a loudness war.

The reason is that virtual instruments are all signal. There's nothing else to them but the foreground sound they're meant to convey. As you pile them up, you don't get a richness of noise. You get a bunch of competing signals.

That lack of noise can extend to live recording in the digital medium, too—even a recording that uses traditional, physical instruments and microphones. In that case, the noise that enters the system may be pretty much the same as in analog recording. But once those sounds are digitized, it's much easier to isolate and manipulate them.

We're all familiar with that process from our digital tools, whatever we use them for. From word processing, to image manipulation, to audio recording, the commands are pretty much the same: highlight, select, and delete.

In digital media, if you can point to something, you can usually eliminate it.

It's even happened to *Pet Sounds*. In 1996, thirty years after its original release, Capitol Records commissioned a digital remix of the album. This time, with the original tapes all transferred to a computer, that conversation in the background of "Here Today" could be isolated...

[POST: "Here Today" conversation isolated]

> Q *Do you have that attached to the flash you have rigged up?*
>
> A *I do.*

And then... it could be deleted. Here's the 1996 digital remix of that same instrumental break. Turn it up as loud as you want, you won't hear those unintended noises in the background any longer.

[BED: Beach Boys, "Here Today" 1996 stereo mix, 1:55ff]

When you choose, as a listener, to focus on what's buried deep in the layers of a recording instead of what's been placed up front to catch your attention—then *you've* changed what is signal, and what is noise.

But if the noise has been eliminated—as digital media can and so often do—then that choice is eliminated as well.

[BED: Galaxie 500, "Tugboat"]

If you've been following this podcast, we've seen examples of this in every episode.

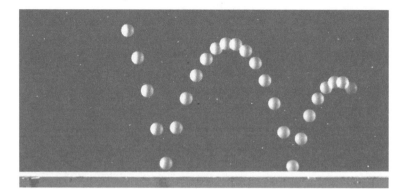

In **episode one**, we heard the difference between real time—the flexible time experienced and expressed by human musicians—and machine time, which is used by computers. Machine time is all signal. But the way a human drummer plays, the spaces between beats are felt as much as the hits. Each emerges out of that background.

⌐¬
∟⌐

In **episode two**, we considered the role of noise in urban space—and the way that digital audio is now allowing us to control the signals we hear in public. Whatever we hear through headphones commands our attention regardless of who or what is around us, cocooning each of us in signal even as we occupy shared space.

And in **episode three**, we compared the quality of expression using all the noise in our voices, against the way cell phones isolate the signal of our words alone. Without all the other sounds around our words—our breath, our proximity, our tone—we lose the fullness of their meaning.

Episode four looked at the way that the internet has altered the terms of exchange for music. But that, too, can be seen as a question of signal and noise. Streaming services strip recorded sound away from its physical format, isolating the signal of music and deleting the noise of its context—not only the material package, but all the information that went along with it: the songwriters, producers, even the names of many musicians themselves.

Finally, in **episode five**, we took up the issue of how these isolated digital signals are being manipulated by online corporations. Or you might even say, how we are each being transformed *into signal* for profitable use by those businesses.

In other words, this question of signal and noise is about more than which mix of *Pet Sounds* you prefer. It has significance far beyond the kinds of arguments music fanatics love to have with each other.

Because the real difference is between a world enriched by noise, and a world that strives toward signal only.

To me, that is the essential change we've been experiencing in this shift from analog to digital communications. I've addressed it here through sound, because that's what I know best as a musician. But also because it's the medium we're sharing now, in this podcast.

Throughout the series, my aim has been to call attention to aspects of sound we may not always think about. You might say, I've been trying to highlight different parts of the noise around us.

And that's because it's my hope that by listening to a wider swath of noise, we might discover more about what is meaningful signal for each of us. And how we might best share those signals with one another.

Ian Coss
Podcasting Glossary

TRACK Scripted narration, read by the host
and typically recorded in a studio.

ACT (short for "actuality") Voices other
than the narrator's, often recorded on location. Also called "cuts" or "sound bites."

SFX (short for "sound effect") An audio
sample that is heard clearly in the foreground and relates to the content of the
story. This might be recorded on location,
sourced from an existing audio library,
or simulated in the studio.

AMBI (short for "ambience") Audio that
creates a sense of location. Also called
"nat[ural] sound."

BED Audio that runs under voices, for a
feeling of continuity or motion. Typically
refers to music, but could be field recordings or other ambient sound.

POST A cue for audio to move from background
to foreground, in order to illustrate a point
or otherwise punctuate the narrative.

Image Credits

Episode 1, **Cover (lower right)**: PHO-FAM001 Familiarization Manual, Mission Control Center Houston, NASA; **Inside front cover**: Courtesy 20/20/20; **5**: Photo: Andrew Davidhazy, courtesy the artist; **6**: Photo: Alessandro Nassiri, licensed under CC BY-SA 4.0; **9**: Courtesy 20/20/20; **12**: Courtesy Ian Coss; **14**: Courtesy Boston Red Sox; **17**: Courtesy Boston Red Sox; **20**: © Thomas Struth; **Episode 2**, **28**: Photo: Laura Knight; **30**: Courtesy Pentagram; **32**: Photo: Carole Teller, courtesy Collection of the Greenwich Village Society for Historic Preservation; **33**: Courtesy Jeremiah's Vanishing New York; **34**: Photo: Neal Boenzi, © Neal Boenzi/*The New York Times*/Redux; **36**: Photo: Matt Weber, © Matt Weber; **38**: © *The New York Times*/Redux; **40**: Smart Destinations, licensed under CC BY-SA 2.0; **47**: Photo: Roberto Masotti, © Roberto Masotti/Lelli e Masotti Archivio; **Episode 3**, **55**: Photo: Stan Wyman, © The LIFE Picture Collection/Getty Images; **57**: Photo: Chris Kalafarski, courtesy PRX; **67**: Vinton Cerf and Robert Kahn, "A Protocol for Packet Network Intercommunication," in *IEEE Transactions on Communications COM*, 1974; **Episode 4**, **71**: Photo: Aylin Güngör, courtesy the artist; **72**: Photo: Minoru Tsuyuki, courtesy 20/20/20; **78**: Photo: Adam Ramet, courtesy the artist and Piano Pavilion; **79**: Photo: Andrew Reed Weller, courtesy the artist; **82**: Photo: Nick Gray, licensed under CC BY-SA 2.0; **84**: Photo: Michael Baca, courtesy Sub Pop; **87**: Photo: Josh Sisk, courtesy the artist; **88**: Photo: Richard Bellia, courtesy 20/20/20; **Episode 5**, **90**: Photo: Amy Gorel, courtesy Amy Gorel/WBUR; **93**: Courtesy Forced Exposure; **95**: Photo by the author; **97**: Courtesy Forced Exposure; **105**: Photo: Caroline Culler, licensed under CC BY-SA 3.0; **106**: Photo: Evan DuCharme, courtesy the artist and *SF Weekly*; **108**: Photo: Ben Roberts, © Ben Roberts/Panos Pictures; **109**: Courtesy City Lights; **Episode 6**, **114**: Photo: Susan Jane Golding, licensed under CC BY 2.0; **119**: © Bruce Eric Kaplan/The New Yorker Collection/The Cartoon Bank; **120**: Photo: André Løyning; **121**: Photo: Ed Bornstein, courtesy Electrical Audio; **127**: © Michael Ochs Archives/Getty Images

Ways of Hearing the podcast is a production
of *Showcase*, from Radiotopia and PRX

Produced by Damon Krukowski, Max Larkin and Ian Coss
Sound design by Ian Coss
Executive producer Julie Shapiro
Thanks to Alex Braunstein and the PRX Podcast Garage

Ways of Hearing the book is a production of MIT Press

Written by Damon Krukowski
Interlude by Emily Thompson
Design and picture research by James Goggin, Practise
Edited by Matthew Browne
Thanks to Janet Rossi and the MIT Press Production Dept.

Compressed audio for **Ways of Hearing** can be streamed
from podcast apps by subscribing to *Showcase*

Ad-free, CD-quality audio for each episode can be down-
loaded from **waysofhearing.bandcamp.com**